VW BEETLE AT WAR

AF734621

Dr. Hans-Georg Mayer

Schiffer Military History
West Chester, PA

Bibliography:

"Automobil-Chronik" magazine, 12/77, 5/79, 8/79

Boschen, Lothar, *Das grosse Buch der Volkswagen-Typen*, Stuttgart, 1983

Etzold, Hans-Rüdifer, *Der Käfer*, Vol.II, 1984

Kubisch, Ulrich, *Allerweltswagen*, Berlin 1986

Oswald, Werner, *Kraftfahrzeuge und Panzer der Reichswehr, Wehrmacht und Bundeswehr*, Stuttgart 1970

Piekalkiewicz, Janusz, *Der VW Kübelwagen Typ 82 im Zweiten Weltkrieg*, Stuttgart 1977

Porsche Archives, Stuttgart

VW Archives, Wolfsburg

Translated from the German by Edward Force.

Copyright © 1992 by Schiffer Publishing Ltd.

All rights reserved. No part of this work may be reproduced or used in any forms or by any means—graphic, electronic or mechanical, including photocopying or information storage and retrieval systems—without written permission from the copyright holder.

Printed in the United States of America.
ISBN: 0-88740-400-6

This title was originally published under the title,
Der VW-Käfer im Kriege,
by Podzun-Pallas Verlag, Friedberg.

We are interested in hearing from authors with book ideas on related topics. We are also looking for good photographs in the military history area. We will copy your photos and credit you should your materials be used in a future Schiffer project.

Published by Schiffer Publishing, Ltd.
1469 Morstein Road
West Chester, Pennsylvania 19380
Please write for a free catalog.
This book may be purchased from the publisher.
Please include $2.95 postage.
Try your bookstore first.

Acknowledgements

For this documentation, pictures have been provided by: Mr. Klaus Parr of the Porsche AG Archives, Stuttgart; Mr. Helmut Pfeifhofer, proprietor of the Porsche Auto Museum, Gmünd, Carinthia, Austria; Mrs. Hackspiel, Mrs. Neefe-Hansmann and Dr. Bernd Wiersch of the VW Archives, Wolfsburg. Hearty thanks to them all.

Cover artwork by Steve Ferguson, Colorado Springs, CO
Additional research by Russell Mueller

RACE YOU TO THE TREE

Two Luftwaffe couriers in their Type 87 Volkswagen Kafer (Beetle) skid to a halt behind the sparse cover of a tree after crossing the path of a marauding RAF Hurricane IIB. Unwilling to brave the run for another tree through the murderous barrage from the Hurricane's twelve .303 guns, the *Unteroffizieren* will have to abandon their unarmed sedan for the long walk to their Tunisian landing field.

The VW is camouflaged in standard blue-grey base paint and over-sprayed with RLM Yellow Sand, which is likewise heavily streaked with dried mud and dust. The *Afrikakorps* palm and swastika emblem is retained on both doors, and the trunk hood against the original base undercoat.

The first series-production Volkswagens in the war year of 1941. The cars have covered headlights for war service.

Volkswagen at War

At the end of 1937 the Volkswagen took on its final form. It was introduced at the laying of the factory cornerstone on Ascension Day, 1938 as the KdF-Wagen (VW Type 60) in three versions: sedan, sedan with opening roof (Kabrio-Limousine) and convertible. Soon other types were developed: The "Beetle" was built individually as an open truck, a station wagon, and even with a streamlined body ("Berlin-Rom-Wagen").

After war broke out in 1939, attention naturally turned to the military utility of the Volkswagen. Initial consultation between the SS-Fahrbereitschaft, which had subjected the KdF-Wagen to long and hard testing, and the Army Weapons office, which was responsible for the provision of military equipment, took place on August 17, 1938. The ensuing developmental work and the thorough testing of models throughout 1939 resulted in the well-known VW Kübelwagen (VW Type 82), of which some 50,000 were built between 1940 and 1945.

Along with the Kübelwagen, the Schwimmwagen amphibious vehicle ranks among the best-known and most cherished VW veterans. Its development goes back to 1940. The so-called long Schwimmwagen (Type 128) was at first built in very small numbers in 1940-41. The final production version (Type 166) went into production in 1942 and was supplied mainly to Waffen-SS divisions.

The Schwimmwagen replaced the earlier motorcycles and sidecars that had proved to be disadvantageous in roadless country.

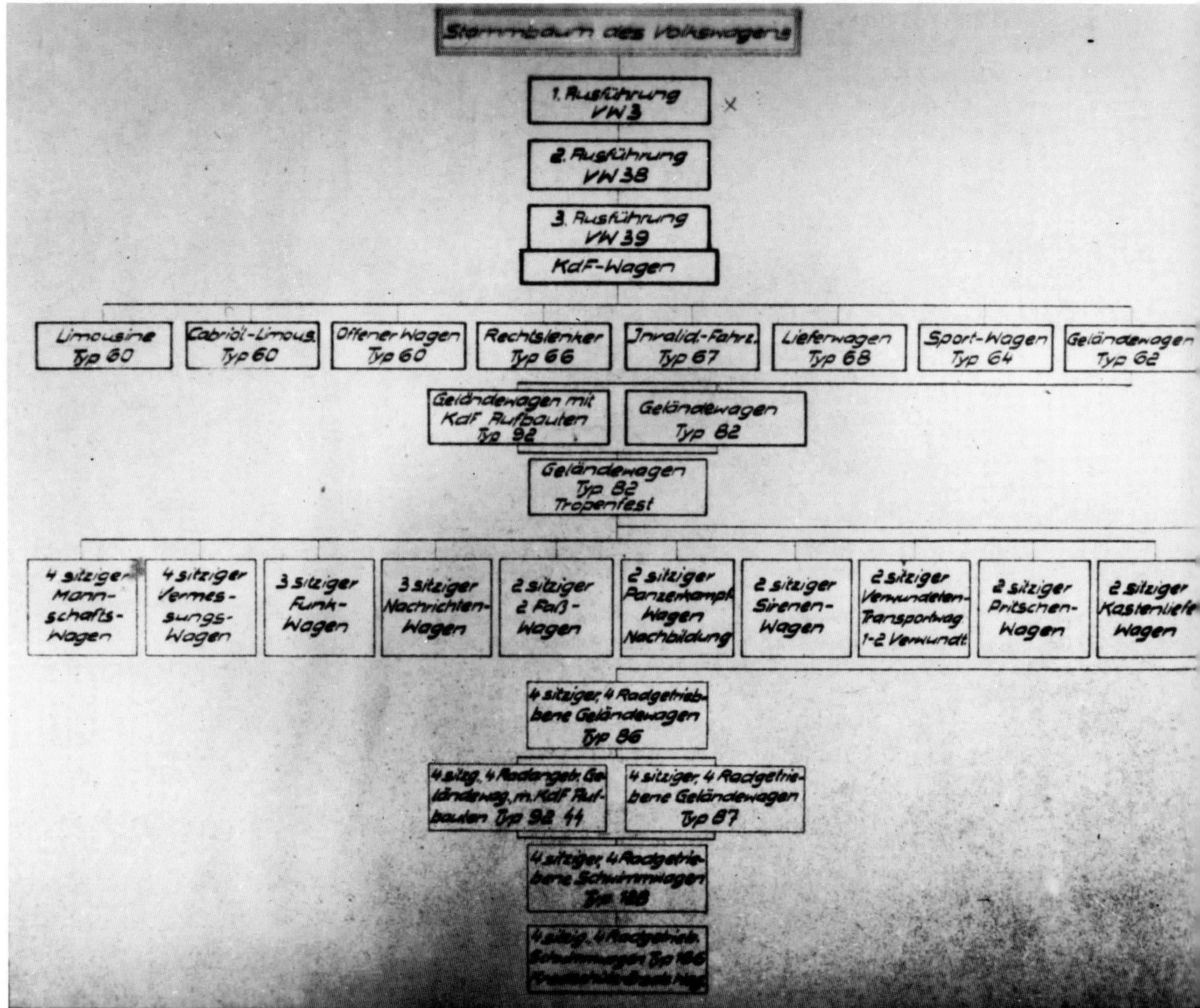

The Volkswagen family tree. It took on its final form in 1938 and thus became ready for series production. This Type 60 was renamed Type 11 after the war and is still built outside Germany today. From the Type 60, using modern kit-building principles, the various military versions of the Volkswagen were developed.

By the end of 1944, when production was halted because of material shortages and rationing, barely 15,000 VW amphibious vehicles had been built.

During the war, production of the civilian KdF sedan was stopped. Along with military vehicles, the Volkswagen factory produced chiefly military equipment, which had priority. The originally planned large-series production of the KdF car had to be postponed until a later date, as yet undetermined.

But production of the VW sedan was not completely given up. When the war began in 1939, 210 Type 60 cars had been produced. On August 15, 1940 the first saleable VW sedan left the new factory in Wolfsburg. Others followed, and in 1941 "the first series-produced Volkswagens" were presented to the public as an act of propaganda.

The authorities wanted to show that, even in a world war, Germany had the capability to turn out luxury goods in great numbers (only in 1942 did the Third Reich convert to a wartime economy).

Despite this relatively effective "public relations" work, only 630 regular KdF cars were produced during the war and supplied to state or party offices.

For military purposes, on the other hand, special versions of the KdF car were developed, namely an off-road sedan based on the Kübelwagen (Type 82 E) and a four-wheel-drive vehicle with the technology of the Schwimmwagen (Type 87). Thus there were three basic models of the KdF Volkswagen:

VW Type 60
VW Type 82 E
VW Type 87

The first KdF cars produced at the Volkswagen Works were formally introduced to the public in 1941. These cars were turned over to state and party offices. Unlike the Wehrmacht Beetles, these cars have gloss paint and chromed parts. The swastika with the gear was the symbol of the German Arbeitsfront (Reich Labor Service), of which the whole KdF organization was a part.

VW Sedan Type 60

The original Volkswagen, the KdF car of 1938, had a bottom platform frame and a central tube backbone, to which an all-steel body was screwed. The running gear consisted of a crank-link front and swing rear axle with individually suspended wheels. The front axle was fastened to the head of the frame, while the rear fork of the central tube carried the drive unit. Power was produced by an air-cooled four-cylinder horizontally opposed rear engine (985 cc, 23.5 HP). The simple sliding-gear or sliding-rod transmission (four forward speeds, one reverse) with integral differential was attached directly to the engine block. The unsynchronized gearbox demanded skill and sensitivity. Anyone who has driven a standard-shift VW before 1965 knows the difficulty of shifting, including double-clutching, from firsthand experience. The old-fashioned spindle steering of the KdF car saw service in the VW 1200 A until 1965 (which also still had the uniform wartime steering wheel with its three spokes). And naturally, all Volkswagens were then equipped with mechanical cable brakes, which remained until 1962.

As a standard rule, all cars had headlight caps with small light slits during the war, as did the KdF car that went into production in 1941. Because of war-related rationing, only vehicles with closed bodies were built. The originally planned Cabrio-Limousine with its big rolling roof panel was not built. The Type 60 was introduced with gloss black finish and chromed bumpers, and looked very impressive from the outside.

As of March 1943, all Volkswagens had a bigger and somewhat more powerful engine (1131 cc, 24.5 HP), which was replaced by a 30 HP engine only in December 1953. The power increase of only one horsepower was modest, to be sure, but it gave this powerplant a markedly better torque.

With the body removed, the running gear of the Volkswagen can be seen very clearly. Front and rear views with engine, unchanged in principle to this day. In 1987, though, only 17,166 Volkswagens were built worldwide.

Above: The Volkswagen Kabrio-Limousine in its original civilian form (Type 60) on the Autobahn near Giessen. The KdF car was built in small numbers as of 1938. This picture was taken on October 4, 1940.

Upper right: The instrument panel of the 1938 KdF Volkswagen. Two glove compartments to the left and right, controls with speedometer, warning lights, and light and wiper switches. To the right of center is an empty panel with a drawing (intended for radio installation). In the middle is the directional switch, under the instruments is the starter button. The simple three-spoke steering wheel remained standard for Volkswagens until 1966.

Right: A KdF delivery truck of 1939 (Type 68).

Upper left: An early convertible of 1939 on the road above Lake Constance. It looks very much like the production model built as of 1949 by Karmann of Osnabrück. The windshield frame also corresponds to that of the VW export convertible of 1949-50.

Above: A 1938 Volkswagen convertible, photographed circa 1940. The covered headlights were required in wartime. Unlike the car shown at left, this one has the windshield and frame of the sedan.

Left: Adolf Hitler in the first VW convertible. This car was kept at his Berghof on the Obersalzberg during the war. After the war it passed through various private hands, and it was bought back by the Volkswagen Works in 1956 for 600 Marks. Today this 1938 Beetle cabriolet is in the auto museum in Wolfsburg. Here the car, photographed in 1940, already has covered headlights.

A KdF car (Type 60) in a new housing project in Wolfsburg (then still called "City of the KdF Car"), 1941-42.

Instrument panel and foot space of the ambulance (Type 67) made in 1939 by rebuilding a regular KdF sedan.

Left: Chassis front — central box frame with gearshift lever and bottom plate; front wheels led by two transverse swing arms, one over the other, and sprung by transverse torsion bars (see #4); individual-wheel steering, fuel tank.

Above: Installing a gas bottle next to the rear seat.

Below: Concealed gas bottle next to the rear seat.

Above: Ducting between the engine and a gas bottle in a VW sedan (Type 240), September 1943.

Kdf-Wagen (VW Type 60), built in 1942-43, with the then-customary matte black paint. Some of these Volkswagens were supplied to the Wehrmacht and reequipped there as needed. These pictures come from VW factory information for a police car.

VW Sedan Type 82 E

A combination of the Kübelwagen chassis and the KdF sedan body appeared in 1941 as the Type 82 E. The chassis component with the Kübelwagen running gear did not differ significantly from that of the civilian Volkswagen, though several changes were made for improved off-road capability. The Kübel chassis was given special front stub axles and transmission (or reduction) gears on the two rear half-axles to increase ground clearance. The Volkswagen on the Kübel chassis looked long-legged in comparison to the usual KdF sedan. The engine and gearbox were essentially those of the 1938 VW. But the rear-axle drive of the wartime Beetle was lowered from that of the KdF car to improve traction. The Type 82 E thus had greater ability to keep going in the lower gears; as a result, the top speeds in each gear and overall were lower (top speed 80 kph). The VW Type 82 E intended for military use had special (5.25-16) off-road tires. This car's equipment was Spartan, limited to what was absolutely necessary.

The cars were delivered in matte black, resprayed by the Wehrmacht to meet their needs or immediately repainted in clear, plain military colors.

The simple tube-frame seats of the Kübelwagen were generally used.

The VW off-road sedan with the normal body of the KdF car and the raised chassis of the Kübelwagen surmounted almost every obstacle, even without four-wheel drive. This type was built from 1941-1946.

The Type 82 E military Beetle had the advantage of an enclosed body over the Kübelwagen. The passengers were thus relatively protected from cold, rain, wind and dust. Depending on how it was used, though, it was much more important to leave the vehicle quickly, and here the narrow two-door Beetle was clearly at a disadvantage. The Beetle bodies were pressed at the Wolfsburg factory, whereas the Kübel- and Schwimmwagen parts had to be obtained from the Presswerk Ambi-Budd in Berlin — one reason why these vehicles no longer could be produced after the war, as Ambi-Budd was in the eastern zone of the capital city of the Reich.

With its more powerful engine and special equipment, some VW Type 82 E cars were supplied to special SS units and designated 92 SS.

Like the Kübelwagen, the Kübel-based Beetle Type 82 E was also equipped with a woodgas generator in times of gasoline shortages. Since gasoline consumption by the Luftwaffe and at the front was especially high, and since Germany had only limited oil reserves in occupied countries (Rumania), in the hinterlands "native fuels" were used. The generator cars are recognizable at first glance by the big bulge in the front hood. The spare tire was then placed on the Beetle's roof, where the firewood or anthracite coal needed to supply power was also kept. Other one-off variations based on the VW 82 E were also made: the pickup truck and a delivery truck (as an early forerunner of the later "Fridolin" light postal service delivery truck).

Right: Production of the off-road sedan. The Kübelwagen chassis and the KdF car body are united. The gears on the rear half-axles can be seen. In the right background is a completed high-bodied Beetle. As of 1941 the off-road Beetle was built for the German Wehrmacht; after the war it was made for the occupying forces as Type 51. On February 2, 1946 the order was given to stop producing the Type 51. All vehicles still under construction were to be converted into the Type 11. This type, though, was probably still produced until October 1946.

Below: A finished off-road sedan leaves the assembly line.

Testing the off-road sedan on the factory grounds. The pictures show the extreme capability of the Volkswagen. Such performance was equaled only by the American Jeep, though with three times the fuel consumption and considerably higher production costs.

An off-road Beetle as an open truck (Type 825), built in 1940-41. Two small doors in the rear and a hatch in the rear bed allowed access to the engine. The rear panel could be unscrewed for easier servicing or removal or installation of an engine. The powerplant, like those of all Volkswagens until 1942, displaced 985 cc and produced 23.5 horsepower.

Left: Type 92 SS: Two rifle racks are attached to the rear sides. Underneath is a first-aid kit. The picture also shows that getting out of the two-door VW sedan quickly in military service was impractical. The Kübelwagen was more advantageous in this respect.

Below: Reclining seats in the Type 92 SS. In the passenger's foot space is an extra gasoline can (with SS runes). This car was built circa 1943.

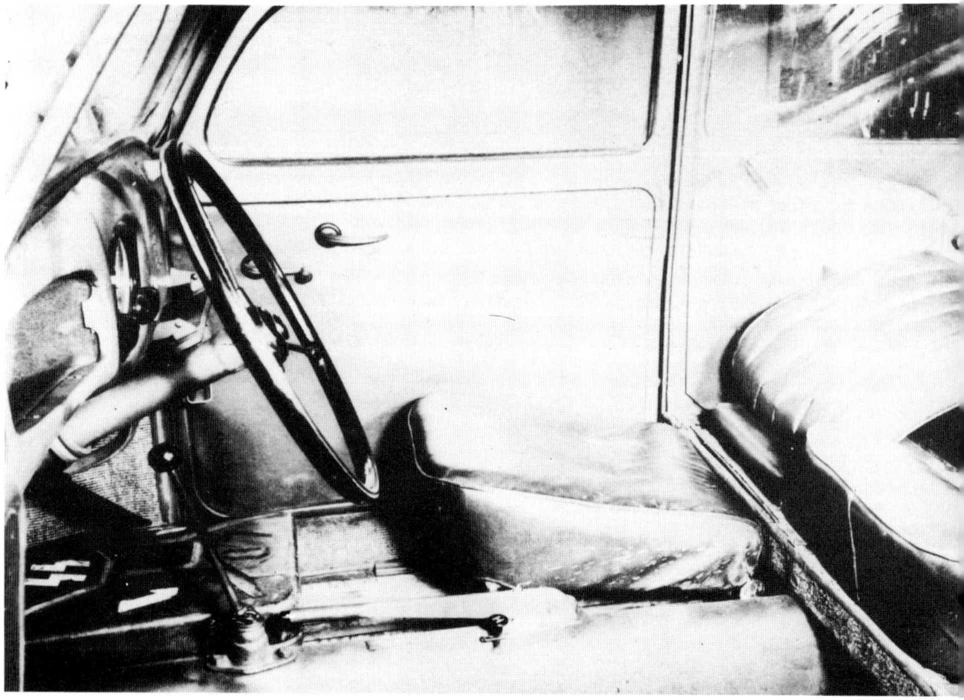

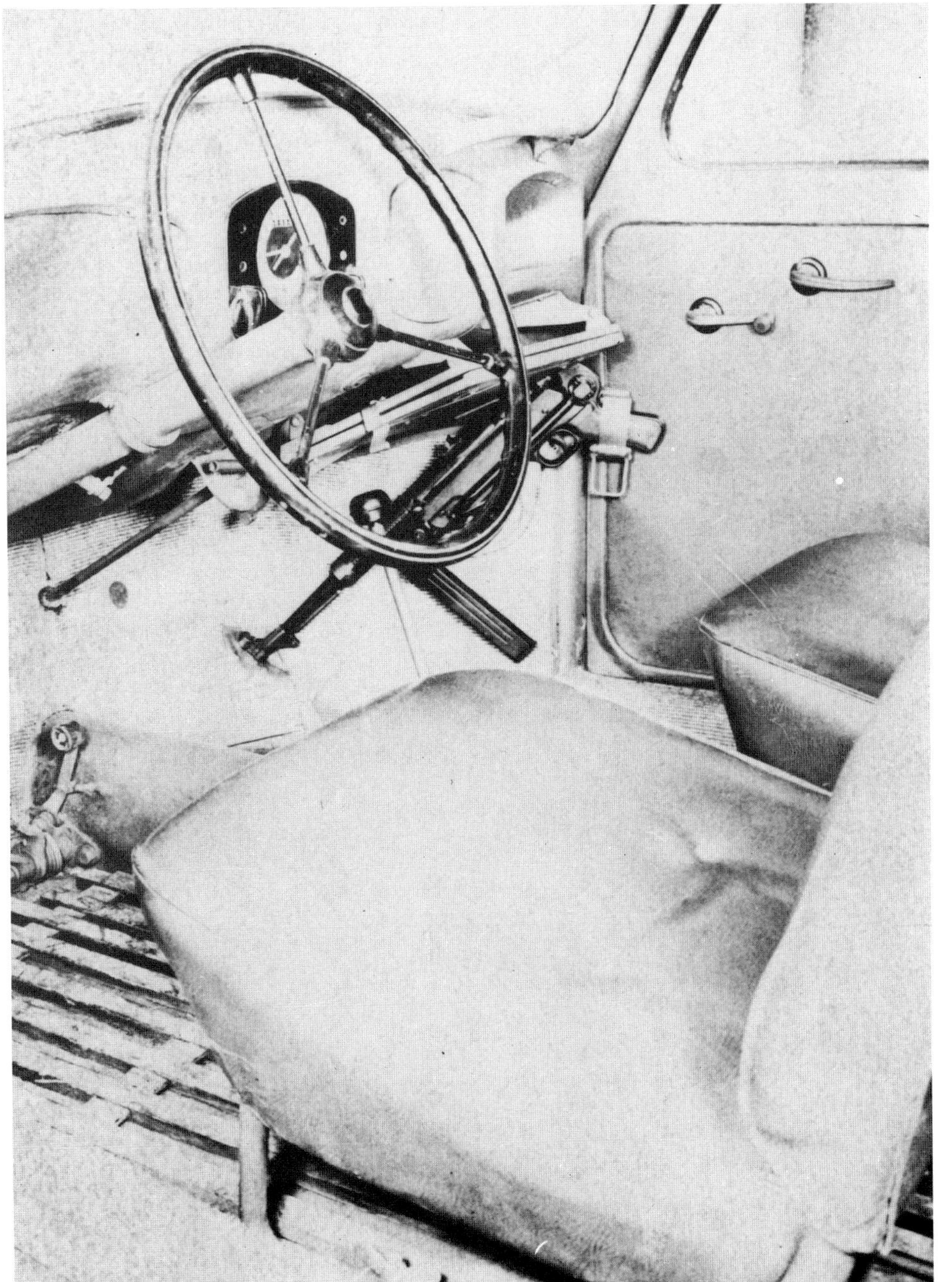

Type 92 SS: Note MP 38/40 machine pistol is attached to the side.

The instrument panel with a sliding desk top in the VW Type 82 E (also in Types 92 SS and 87). The machine pistol can be seen in the right foot space. Wooden bars like those used in the Kübelwagen form the floor. The simple attachment of the seats with wing nuts was used in the standard VW until 1965.

The VW box van (Type 81) being tested on the Autobahn in May of 1941. This is the so-called tropical vehicle on the Kübelwagen chassis. The rear view with opened door shows the engine under the rear bed. The carburetor was moved to the side to gain load space. The Type 81 was a forerunner of the VW light delivery van that was built from 1964-1974, mainly for postal use.

In the summer of 1941, two Volkswagen 82 E Beetles in special form were prepared for the Office of Colonial Policies, which sent them on a trip to Afghanistan. Such far-flung positions in the Orient, along with the uprisings of Arab nationalists against British colonial governments, were of importance in Germany's battle against British positions in the eastern Mediterranean area and in India. The little convoy, which was to make contact with a German outpost of the Office of Colonial Policies, consisted of one truck loaded with a variety of military equipment, various cars, a VW Kübelwagen and the two KdF 82 E Volkswagens. The trip went from Berlin through the Balkans to the Black Sea. After traveling by sea from Istanbul to Trabzon, the column passed through Iran to Teheran and then to the Iranian-Afghani border. It reached its destination after four weeks, though the trip was very difficult. Serious problems were caused by dust in the engines (carburetors, oil coolers, valves, fans, and cylinder-head ribs).

There was also constant trouble with the tires. Because of the high ground temperatures, the temperature of the tires naturally rose too. The tire pressure could double, and even at normal speeds, the rubber surface came loose. The two Beetles must have arrived in Kabul somewhat the worse for wear. Their return trip to Germany was made impossible when Persia, friendly to Germany, was occupied by British and Russian troops on August 25, 1941. On instructions from the Todt Organization, the Volkswagens were apparently destroyed on the spot by their drivers.

During World War II, the German embassy in Kabul, Afghanistan took on added importance to German policies and strategy. Foreign Minister von Ribbentrop insistently urged the Afghani government to collaborate with Germany. The purpose was to support Islamic freedom fighters against British and Russian colonial regimes. In June 1942, a column of vehicles started out from Berlin to support the German position in Afghanistan. Two VW sedans were taken along.

In 1941 two off-road sedans (Type 82 E) in non-military form were built for the Office of Colonial Policies. Special equipment included high-gloss paint, chromed bumpers and handles, covered and concealed headlights, klaxon horn. The vehicles bore Berlin license plates.

Right: The dashboard of the Afghan Beetle.

Below: As early as 1941, the Afghan Beetle had an auxiliary air intake in front of the windshield.

Lower right: The Afghan Beetle had a short-wave radio. Note that the bulkhead between the interior and the engine is perforated with wide louvers for better engine cooling in tropical heat.

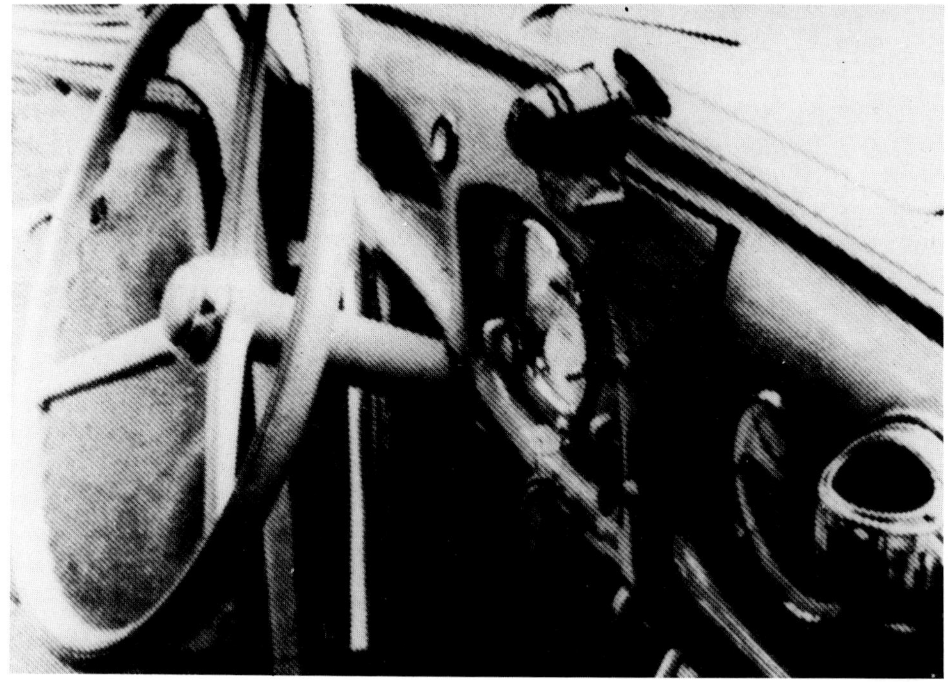

Above: The two Volkswagens in Berlin in 1941, before they started out for Afghanistan.
Below: The column of German vehicles makes a stop in an East European city. Behind the two off-road Volkswagens is a VW Kübelwagen; at right is an Adler Diplomat.

Left: The German convoy to Afghanistan started from Berlin on May 27, 1941, shortly before the Russian campaign began. The first ports of call were Dresden, Prague, Vienna, Budapest, Bucharest and Constanza on the Black Sea, where the vehicles were loaded onto a Rumanian freighter. The picture shows the two VW Beetles on the Autobahn during their drive southward. In the Office of Colonial Policies it was intended to equip civil administrations with tropical versions of the off-road Volkswagen 82 E, 87 and 166 in the future.

Right: An off-road Beetle Type 82 E (one completely original) in the Technical Museum at Sinsheim. In the background is a VW Kübelwagen.

Because of the fuel shortage toward the end of the war, many vehicles were converted to run on woodgas. These two Wehrmacht Volkswagens are generator-powered, hence the big bulges in the front hoods. At left is a VW Kübelwagen, at right a VW off-road sedan.

Design for the rebuilding of a VW Type 92 off-road sedan to take woodgas drive (Type 230). On the front end of the car is the generator. The spare tire is attached to the roof. All generator-powered vehicles had a roof rack on which the fuel was carried.

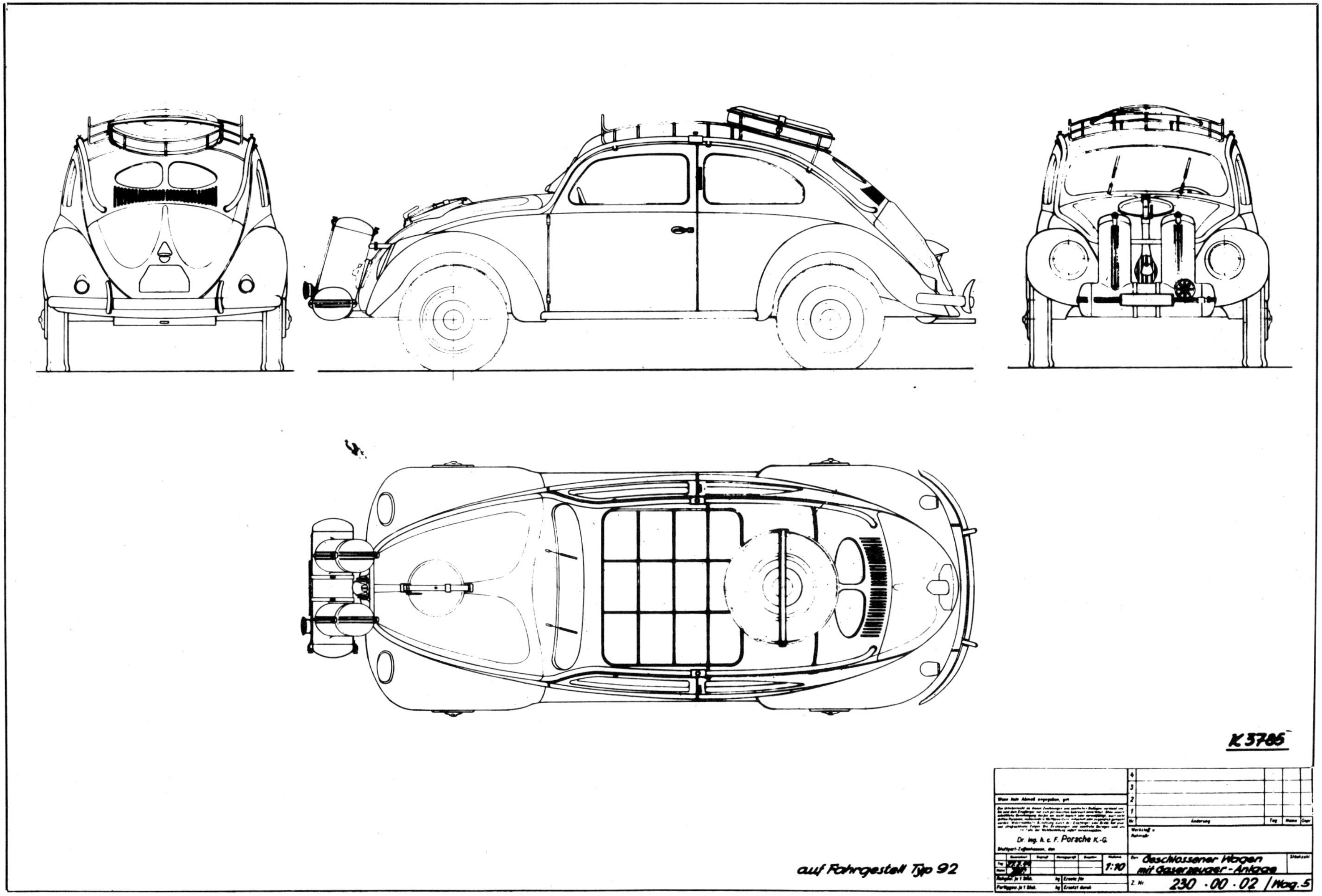

Right: A KdF car on the raised chassis of the Kübelwagen, with a woodgas generator (Type 230/4, 1943). The generator rises well above the nose of the car. The front hood was braced.

Below: An off-road sedan converted to run on woodgas, September 1943.

Lower right: Here the "refueling" of a VW off-road sedan with generator drive is being demonstrated.

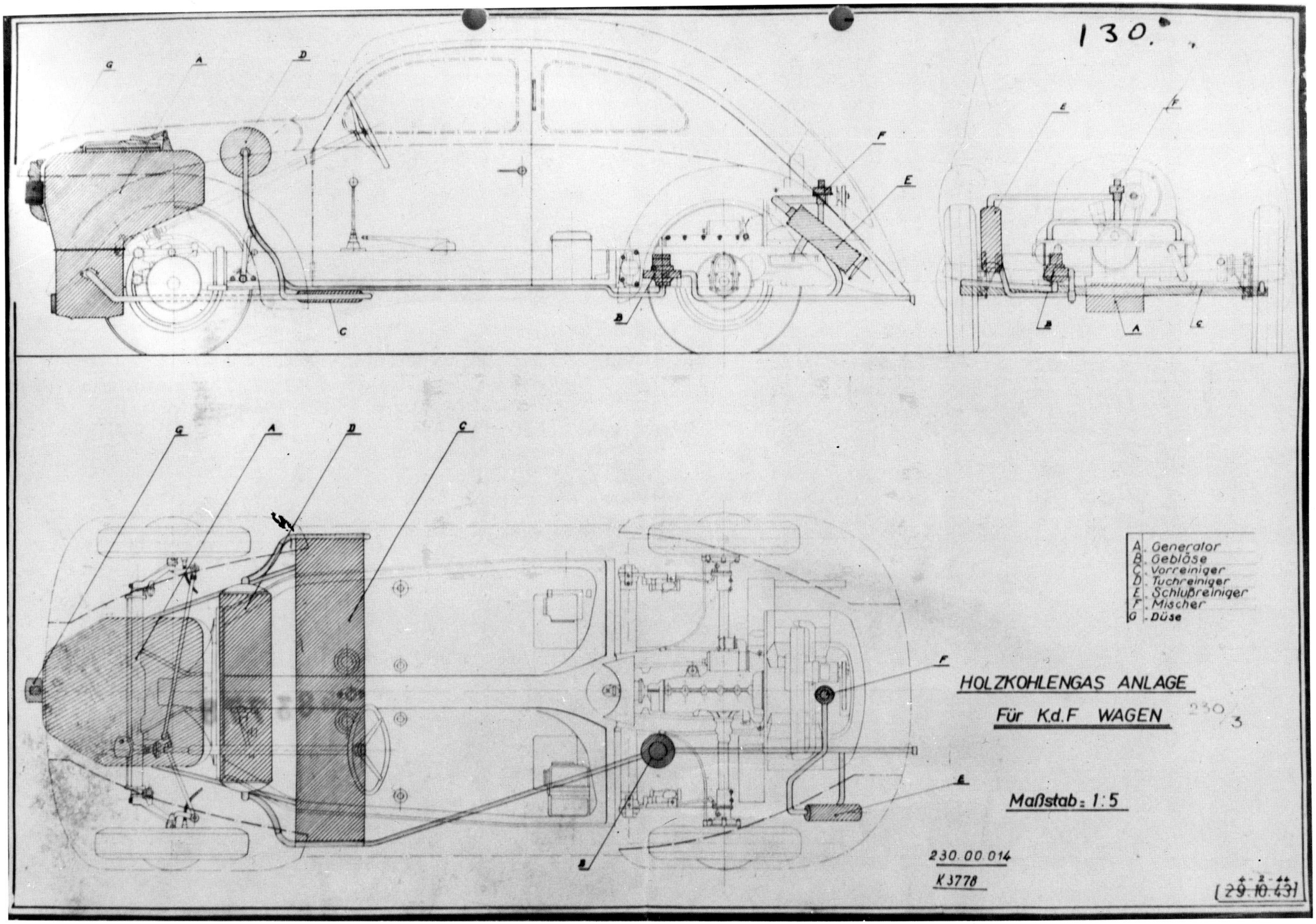

The woodgas system for the KdF-Wagen, Porsche Type 230/3, installed in a VW Type 82 E, 1944. (A. Generator, B. Fan, C. Filter, D. Screen, E. Final Filter, F. Mixer, G. Nozzle)

The opened hood of the lengthened nose gives a view of the generator and first filter.

Above: Off-road sedan with woodgas system (Type 230). The generator is covered by a special front hood. With the low gear ratios of the Kübelwagen and snow chains, this vehicle is quite capable of being used in extreme winter weather. Built in 1944.

Below: A VW off-road sedan (Type 82 E) with a built-in woodgas system. This unusual vehicle was used by the Porsche firm itself in 1944.

Front view of an off-road sedan with woodgas generator, built in 1944. The front hood is held by tension springs and has been cut open to make room for the big generator. In the background is a factory yard covered with camouflage nets.

With the hood open, one can see the generator and first strainer. The generator rises well above the nose of the car, which limited the off-road mobility of the Wehrmacht VW to some extent.

VW Sedan Type 87

Naturally, attempts were soon made to make the VW design, already capable of off-road use, even better by adding four-wheel drive.

On July 1, 1940, Ferdinand Porsche AG was commissioned by the Army Weapons Office to develop an amphibious off-road car. The first prototype was finished in September of the same year. Parallel to this were attempts to adapt the four-wheel drive of the Schwimmwagen for use in the Kübelwagen (Type 86) and the KdF car (Type 87). The VW 86 did not get farther than a study. Only two vehicles were actually built. On the other hand, the four-wheel-drive Beetle was built in a small series between 1941 and 1944. The VW 87, the "commander's car", combined elements of various VW types. The chassis came in principle from the KdF or Kübelwagen. The central tunnel, though, was bigger to hold the necessary driveshaft. The front and rear axles were also built essentially the same. Naturally the Type 87, like the other military versions of the Volkswagen, also had additional gears and production differential locks (here on both the front and rear axles). The front end had special stub axles, which were needed to connect with the driveshafts. In addition, the VW 87 had the simple final drive, though with an additional off-road gear like that of the Schwimmwagen, which could be engaged like the front-wheel drive, by using a shift lever in the middle of the car. Because of this, the hand-brake lever also had to be moved.

A look under the front hood of a VW Type 87. The fenders of the four-wheel-drive Beetle were widened by having a strip of sheet metal added where they met the body. This can be seen clearly in the picture by the joint next to the headlight. It was necessary to widen the fenders to allow the huge off-road wheels to be mounted. This vehicle, though, has the narrower normal tires of the Kübelwagen and the Type 82 E.

A Type 87 training chassis is in the Porsche Museum at Gmünd, Carinthia. The half-shafts of the front axle, the divided right track rod and the brake cables can be seen. A second shift lever for the optional four-wheel drive is mounted on the frame tunnel. For reasons of space, the hand brake thus had to be moved farther forward and be mounted in a vertical position to be easier to reach. The chassis ran on Kronprinz sand tires, as did the cars of the *Afrika-Korps*. The tropical air filter is also the same.

The powerplant of the VW 87 was the stronger VW engine with 1131 cc displacement and 24.5 horsepower. Wheels and tires for the Schwimmwagen and Commander's Car were a special type (5.25-16 or 200-16 tire size), which also necessitated changes in the bodywork. The fenders (and thus also the running boards) of the VW 87 were wider to accommodate the big off-road tires. The VW 87 also had a big opening roof, unlike its closed sister, the VW 82 E. Since many Commanders' Cars were used by the *Afrika-Korps,* they were also made in tropical form, as were some Kübelwagens, with extra protective measures against dust in the electric and fuel lines. The tropical versions also had special sand tires and "Kronprinz" wheels.

Test drives were held in Yugoslavia, Bulgaria, Rumania and Greece in August and September of 1941, to test the suitability of all possible motor vehicles of the German Army for tropical conditions.

Five Volkswagens were included in these tests: a KdF sedan (Type 60), two Commanders' Cars (Type 87), and two Kübelwagens (Type 82). In an exhaustive report by Herbert Kees on November 12, 1941, all the test results were examined and improvements suggested. Experience with the Kübelwagen in the North African theater of war could also be evaluated. The summarized results were published in the spring of 1942 in a pamphlet on the use of motor vehicles in the desert.

A look under the front hood of the 1943 Type 87 Commander's Car: an extra fuel can is stowed in the spare-wheel well because the big off-road tire will not fit in it. A slightly adapted fuel tank with an angled filler was used to make room for the big spare wheel.

VW off-road sedan with four-wheel drive (Type 87). The car has widened fenders and running boards to make room for the big off-road tires. The production Type 87 had a big folding top. From 1941 to 1944, 564 of these cars were built; most of them were supplied to the *Afrika-Korps*.

A restored four-wheel-drive Beetle (VW Type 87).

Land:	Deutschland
Hersteller:	Dr. Ing. h. c. F. Porsche KG und Volkswagenwerk GmbH Wolfsburg
Typ:	Wehrmachtsbezeichnung le. gl. PKW, 4 sitziger,
Preis:	4-Radgetriebener Geländewagen
	Typ 87

Herstellungsjahr:	1941-44
Informationsgrundlage:	Typenbeschreibung vom 27.11.41
Die in ()gesetzten Daten beziehen sich auf Typ:	
Bemerkungen:	nur in geringen Stückzahl hergestellt

Motorhersteller:	Volkswagenwerk Typ 1
Zylinder:	4 Boxer
Bohrung:	75 mm
Hub:	64 mm
Hubraum:	1131 ccm
Kompression:	5,8
U/min:	3000
Leistung:	25 PS
Leistungsgewicht:	
Anordnung der Ventile:	hängend
Kurbelwellenlager:	4
Vergaser: Anzahl	1
Typ	Solex 26 VFJ
Zündfolge:	1-4-3-2

Anlasser:	Bosch EED 0,4/6
Lichtmaschine:	Bosch REDK 130/6/2600
Batterie: Anzahl	1 Volt: 6 Amp.: 75
Kraftstoffanlage:	Pumpe
Kühlung:	Luft, Gebläse
Kupplung:	Einsch. tr. F. & S. K10
Getriebe:	Dr. Porsche KG, Schub-
Anzahl der Gänge:	5 V. 1 R.
Antrieb:	4-Rad
Übersetzung Hinterachsantrieb:	v. 5,31 h. 7,31
Höchstgeschwindigkeit:	80 km/h
Lenkung:	Schrauben-Bauart, Porsche
Wendekreis:	10 m
1. Gang:	2. Gang
3. Gang:	4. Gang

Federung: vorn hinten	Drehstäbe
Schmierung des Fahrgestells:	Hochdruck
Bremsen: System Art	mechanisch Innenbacken
Fußbremse-Wirkungsweise:	4 Räder
Handbremse-Wirkungsweise:	4 Räder
Reifengröße: vorn hinten	5,25 - 16
Spur: vorn	1356
hinten	1360

Radstand:	2400
Bodenfreiheit:	255 mm
Länge über alles:	3830 mm
Breite über alles:	1620 mm
Höhe über alles:	1720 mm +
Gewicht des Fahrgestells:	
Gesamtgewicht:	zul. 1240 kg
Nutzlast:	450 kg
Anzahl der Sitzplätze:	3 oder 4
Tankinhalt:	
Kraftstoffverbrauch:	8,5 l
Ölverbrauch:	0,12 l

+ = offene Ausführung - Kraftstoffvorrat: 40 l

The 1945-1946 Volkswagen Types

By the end of the war, the greater part of the Volkswagen factory had been destroyed by Allied air raids.

In the buildings that were only slightly damaged, military vehicles of the British occupying forces were being repaired. Under the greatest difficulties, with the lack of skilled labor, materials and energy of particular seriousness, vehicles for the officials and occupying powers began to be built again in the next months. Removed machinery and equipment was brought back under the direction of the British military government by autumn. Despite all problems, 1785 vehicles could be built out of remaining parts by the year's end, including 977 high-wheel Volkswagens with the body of the KdF car on Kübelwagen chassis. The old Type 82 E, now designated Type 51, was supplied in goodly numbers (about 1800) to the occupying armies. Along with this closed off-road sedan there was also a version with an opening roof (Type 53). For the first time, a convertible was also built on the Kübelwagen chassis, so that for a very short time there was a quasi-military high-wheel variation of the later so popular Beetle convertible (Type 55). On the basis of the Kübelwagen chassis, the Beetle with box body was built as a mail truck and ambulance (Type 83), as well as a shortened towing tractor (Type 100). All these vehicles were variations of the VW Type 82 E, which went into production in 1941. As of October 1945, the Beetle with its normal chassis also went back into production (as Type 11, formerly Type 60). Along with the Beetle sedan, a few Kübel-

The last Kübelwagen for the German Wehrmacht was produced on April 9, 1945. On April 12 the "City of the KdF Car" was surrendered to the Americans. On June 3 it was taken over by the British. The barrack city was overflowing with prisoners of war of all nationalities, German refugees from the east, and foreign workers. Looting and acts of violence took place. The German population fled into the nearby forests. The military government took the position that the factory could be used as a repair shop for American and British military vehicles. In chaotic conditions, though, automobile production was commenced again. Production after May 1945: 1785 vehicles, 977 of them with the KdF sedan body, the rest Kübelwagens.

wagens (new type number 21) were built out of remaining parts — and now also with open and box truck bodies (Types 27 and 28). Thus several versions of the Volkswagen — mainly the Type 82 E/51 — were still produced for a time after the war. The last high-wheel Beetle left the still-temporary assembly line in October of 1946. In Wolfsburg, work was now concentrated more strongly on the peacetime vehicle, the original KdF car. And as of 1949 Heinrich Nordhoff began the continual, years-long improvement of the Volkswagen, which is still built overseas to this day.

In the summer of 1945, Ivan Hirst (in photo, with glasses) was appointed factory commander by the British occupying forces. The British Army demanded further production of Volkswagens to meet the most urgent transport needs. But there was also concern about the spread of Communist agitation, for what with the catastrophic conditions in the factory and the city, and the closeness of the Russian zone, there was great danger that they could also overrun the western zones. On October 22, 1945 the first civilian postwar Type 11, ex-Type 60 sedan was produced and turned over to factory director Rudolf Brörmann (the civilian with hat in the photo). The VW convertible that is at the disposal of the British Army here was, as is easy to see, still a rather crude vehicle. Even the rear window was dispensed with.

A row of Type 51 off-road Beetles is turned over to the British occupation troops. By the end of 1945, the Kübelwagen and off-road Beetle were practically the only types produced, and served to improve the British forces' supply of motor vehicles.

In March of 1946 the 1000th postwar VW rolled off the temporary assembly line. In that year the prices went up and down several times. Until March a Volkswagen cost 4150 RM, in April 5000 RM, in July only 4000 RM, in December once again 5000 RM. Thus a correct calculation was not yet possible. To be sure, one could only buy them for occupation currency, which scarcely anyone had. Most of the cars were given to the occupation troops as reparations, a few to the German postal service or other German agencies. At that time the VW Beetle still had Kübelwagen headlights because of material shortages, and until 1948 only matte colors because better paint was lacking. Hubcaps were also dispensed with.

Below: CCG vehicles (CCG = Control Commission for Germany) wait to be transported to the Allied occupation forces.

An off-road Beetle with four-wheel drive and a front tube bumper. The fenders were adapted according to use. The production of this vehicle for some time after the war would have made sense. There certainly would have been enough customers in the mountainous regions and developing countries. The Bundeswehr or other armies would also have been interested. The vehicle shown here was built, along with a second one, for testing purposes in November of 1946, and is now at the factory museum in Wolfsburg. The second car was used by the French Army and unfortunately lost. The tube bumper was adopted from British military vehicles.

Above: The interior of the four-wheel-drive Beetle is much like that of the KdF sedan. Between the seats is the shift lever for four-speed four-wheel drive.

Below: A four-wheel-drive off-road Beetle of 1946. The side exhaust pipes come from the Kübelwagen; the ends of the tailpipes come out the lower corners of the fenders. This design avoided damage to the exhaust pipes in off-road driving.

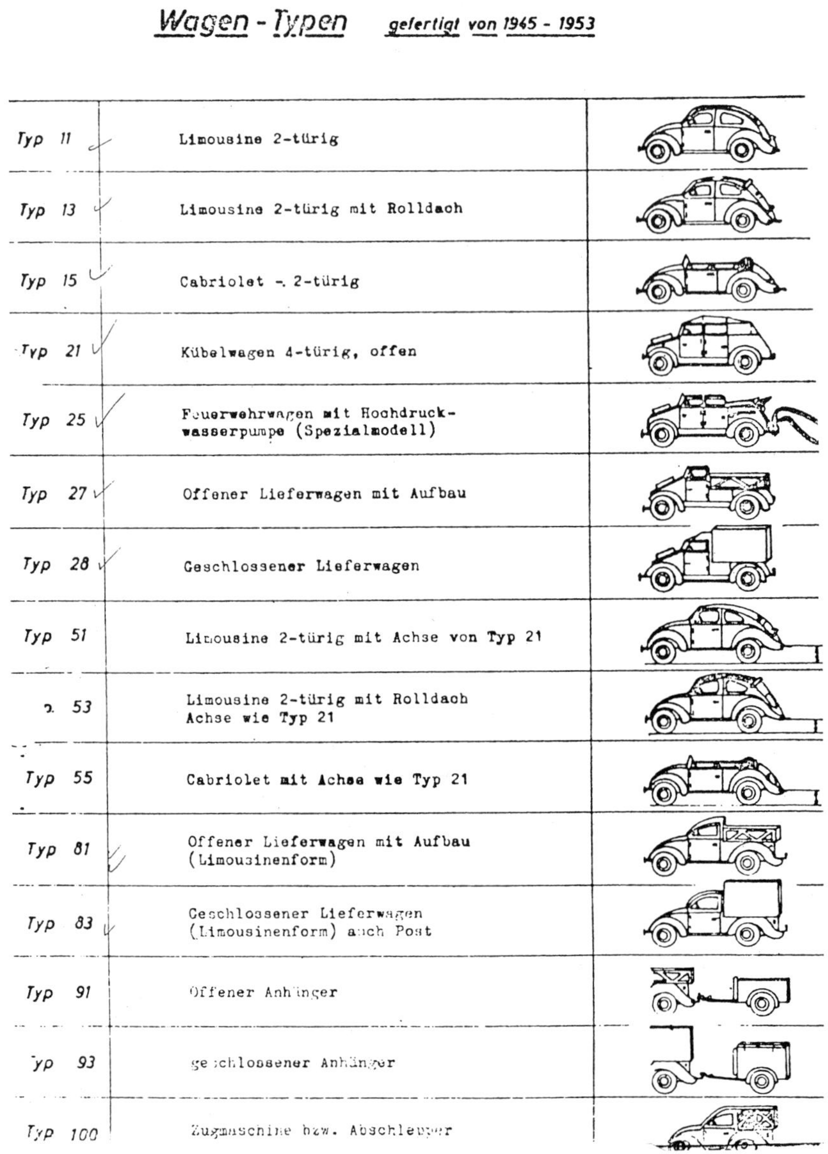

Wagen - Typen gefertigt von 1945 - 1953

Typ 11	Limousine 2-türig	
Typ 13	Limousine 2-türig mit Rolldach	
Typ 15	Cabriolet -. 2-türig	
Typ 21	Kübelwagen 4-türig, offen	
Typ 25	Feuerwehrwagen mit Hochdruck-wasserpumpe (Spezialmodell)	
Typ 27	Offener Lieferwagen mit Aufbau	
Typ 28	Geschlossener Lieferwagen	
Typ 51	Limousine 2-türig mit Achse von Typ 21	
Typ 53	Limousine 2-türig mit Rolldach Achse wie Typ 21	
Typ 55	Cabriolet mit Achse wie Typ 21	
Typ 81	Offener Lieferwagen mit Aufbau (Limousinenform)	
Typ 83	Geschlossener Lieferwagen (Limousinenform) auch Post	
Typ 91	Offener Anhänger	
Typ 93	geschlossener Anhänger	
Typ 100	Zugmaschine bzw. Abschlepper	

A list made under British direction in 1945, showing VW types in production. Off-road military vehicles, such as had been used already by the Wehrmacht, were chiefly built in 1945-46.

The Beetle Type 100 was a stark, very Spartan vehicle. It had the simple seats of the Kübelwagen and the Wehrmacht off-road sedan. The doors were not covered. Cranked windows also had to be eliminated to save material. But in those needy times, it was very useful and more economical than most towing tractors.

Above: A former off-road sedan (Type 51, built until 1946) has been rebuilt into a light semi-trailer cab and used to transport broken-down cars.

Below: The VW Beetle as a short-chassis towing tractor (Type 100) was made in 1945 in very small numbers. It was technically much like the military off-road VW. The low gear ratios were very useful for towing heavy loads. It was also used by the British and French forces.

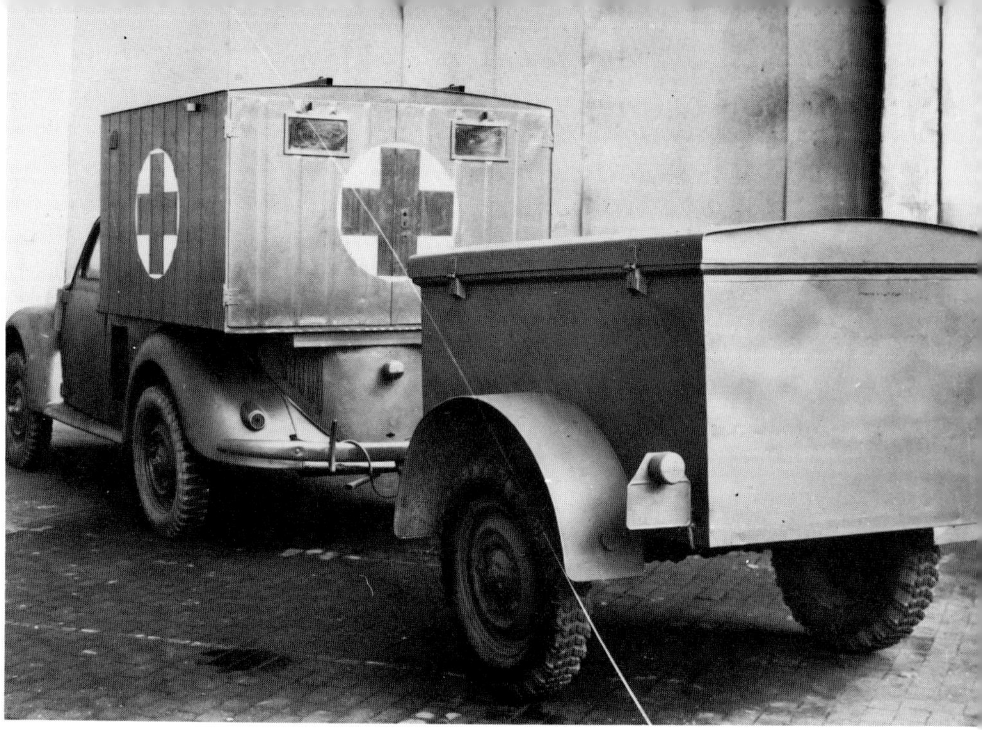

Upper left: This VW ambulance was built in 1945, based on the off-road sedan. The chassis, including the wheels and tires, was taken straight from the Kübelwagen. So were the headlights, as the normal VW headlights were no longer available at the war's end because of material shortages.

Above: The makeshift ambulance with trailer. Such vehicles were also supplied to the postal service. The trailer was built by VW (Type 93).

Left: The needs of the time resulted in unusual vehicles, as a look inside the ambulance shows. The patient's feet stick into the driver's compartment because the box body is too short for the stretcher. In this dark box, even a short trip must have been agony on account of the very hard suspension.

CCG vehicles are picked up, 1946-47. The last inspection before shipping is underway. These are normal Volkswagens (Type 11, ex-Type 60), later the standard sedan. The British forces often used the German Volkswagen as a service vehicle because their own vehicles had suffered from use in the war.

The occupation forces ruled in defeated Germany. The British Army preferred to use the German Volkswagen, because the VW factory was in the British zone. The picture shows a British officer in a VW convertible, circa 1946.

Volkswagens made for the French occupation forces are shipped out by rail, summer 1946.

Wrecked VW's of the British occupation troops.

The VW sedan in its simplest form was also used in great numbers by the German Bundeswehr until 1985. In the last years, of course, all VW Beetles for the German market were imported from the Mexican VW branch factory in Puebla. The photos show a Bundeswehr Beetle from the last German production of 1977.

Little has changed inside: There is still a sheet-metal dashboard with central instruments and open glove compartments, a simple steering wheel, primitive door coverings, and an old-fashioned reserve valve on the floor instead of a more modern gas gauge.

Technical Data

	Type 60 KdF Sedan	Type 82 E Beetle on Kübel- Chassis	Type 87 4WD Beetle
Engine	985 cc, 23.5 HP, as of March 1943 1131 cc, 24.5 HP		1131 cc, 24.5 HP
Gearbox	4-speed	4-speed (special gear ratios)	4-speed + off-road gears (special ratios)
Drive	Rear-wheel	Rear-wheel	Four-wheel
Differential	–	rear	Front & rear
Final drive gears	–	on rear half-shafts	
Ground clearance (empty/loaded)	220/200 mm	290/275 mm	265/245 mm
Tires	4.50-16	5.25-16 off-road	5.25-16 off-road or 200-16; 200-12 for tropics
Fuel tank	25 liters	40 liters	40 liters
Top speed	100 kph	80 kph	80 kph
Years built	1938-40 few, 1941-45 small series	1941-46	1941-44
Numbers made	1938-40: 210 1941-45: 630	Type 82 E: 546 Type 51-53: ca. 1800	564 (2 in November 1946)

Also available from Schiffer Military History

OPERATION BARBAROSSA
OSTFRONT 1944
ROMMEL IN THE DESERT
PANZERKORPS GROßDEUTSCHLAND
THE HEAVY FLAK GUNS
THE WAFFEN SS
THE 1st PANZER DIVISION
THE GERMAN INFANTRY HANDBOOK